Lessons on Poultry For Rural Schools

by US Dept. of Agriculture

with an introduction by Jackson Chambers

This work contains material that was originally published in 1916.

This publication is within the Public Domain.

This edition is reprinted for educational purposes
and in accordance with all applicable Federal Laws.

Introduction Copyright 2017 by Jackson Chambers

IMPORTANT NOTE & DISCLAIMER

IMPORTANT NOTE :
As with all reprinted books of this age that are intended to perfectly reproduce the original edition, considerable pains and effort had to be undertaken to correct fading and sometimes outright damage to existing proofs of this title. At times, this task can be quite monumental, requiring an almost total rebuilding of some pages from digital proofs of multiple copies. Despite this, imperfections still sometimes exist in the final proof and may detract slightly from the visual appearance of the text.

DISCLAIMER :
Due to the age of this book, some methods or practices may have been deemed unsafe or unacceptable in the interim years. In utilizing the information herein, you do so at your own risk. We republish antiquarian books with no judgment or revisionism, solely for their historical and cultural importance, and for educational purposes.

Self Reliance Books

Get more historic titles on animal and stock breeding, gardening and old fashioned skills by visiting us at:

http://selfreliancebooks.blogspot.com/

INTRODUCTION

I am very pleased to present to you another useful poultry title – **Lessons On Poultry for Rural Schools**. It was first published in 1916 by the *U.S. Department of Agriculture*, and is also known as **Farmers' Bulletin No. 464**.

Small scale farming and back-yard farming is in a renaissance right now. More and more people are taking notice of the unhealthy and artificial chemicals and concoctions that are being added to our food and drink. Some have come to the conclusion – myself included – that the only way to be sure you are eating clean, healthy food, is to raise it or grow it yourself.

It's wonderful to see so many schools all over the country with their own Organic Garden growing projects, where the kids take so much pride in the fruits and vegetables that they have lovingly planted, tended and harvested.

This book could be the natural progression from that wonderful veggie garden – raising their own chickens for eggs.

Featuring chapters on *Selecting the Flock, Disposing of Culls, Poultry Feeding, Winter Management of Poultry, Egg Selection, Brooding of Chickens, Raising Crops for Poultry,* and more.

Lessons On Poultry for Rural Schools is a fast read, with a lot of information packed into its 42 pages. Even although it was written with school projects in mind, it is a good starting point for anybody embarking on raising poultry, or those considering it.

Jackson Chambers,
State of Jefferson, November 2017

UNITED STATES DEPARTMENT OF AGRICULTURE

BULLETIN No. 464

Contribution from the States Relations Service
A. C. TRUE, Director.

Washington, D. C.   **PROFESSIONAL PAPER**   December 30, 1916

LESSONS ON POULTRY FOR RURAL SCHOOLS.[1]

By F. E. HEALD, *Specialist in Agricultural Education*.[2]

CONTENTS.

	Page.		Page.
Introduction	1	Incubation	18
Selecting the flock	8	Marketing eggs	19
Disposing of culls	10	Brooding of chickens	20
Poultry houses and yards	11	Preserving eggs	21
Poultry feeding	12	Raising crops for poultry	22
Winter management of poultry	15	Summer management of poultry	23
Poultry diseases and pests	16	Supplement	25
Egg selection	17		

INTRODUCTION.

In the rural-school work in agriculture there is probably no subject which applies more of the principles of animal husbandry than that of poultry culture. This subject provides good material for instruction, is within the grasp of elementary pupils, and the home application may be made with profit in every community and by nearly every pupil.

The points of superiority of poultry study from the standpoint of home application have been arranged in chart form in figures 1 and 2. To be educational in a maximum degree the treatment of this subject should provide at least three features. There should be (1) school lessons arranged in a seasonal sequence, reenforced by (2) practical exercises in or near the school, and applied by the pupil at home in (3) a project which may or may not be club work.

---

[1] Prepared under the direction of C. H. Lane, Chief Specialist in Agricultural Education, States Relations Service.

[2] The writer is indebted to R. R. Slocum, of the Bureau of Animal Industry, for valuable assistance.

NOTE.—This bulletin is intended to assist rural school-teachers to utilize the available publications on poultry husbandry in seasonal school lessons and related home practice during the school year.

61395°—Bull. 464—16——1

THE PROJECT.[1]

Every pupil who has not already a home project or a club project should be induced to begin in September a project with a flock of hens or pullets, the size and scope of the project depending upon the maturity of the pupil, on home conditions, and on parental cooperation. One of the following cases may be met:

(1) Purchase of a new flock by the pupil.
(2) Flock previously reared by the pupil, possibly as a club member.
(3) A part of the parents' flock selected and ownership transferred to the pupil.

POULTRY HOME PROJECT.

ADVANTAGES AS SCHOOL WORK:

1. *INVOLVES ALL PHASES OF ANIMAL STUDY: BREEDING, HOUSING, FEEDING, SANITATION.*
2. *LIFE CYCLE OF HEN NOT LONG.*
3. *DAILY ACCOUNTING REQUIRED.*
4. *IMPROVEMENT RELATIVELY EASY.*
5. *EASILY COORDINATED WITH PLANT PROJECTS.*
6. *CORRELATES READILY WITH OTHER BRANCHES.*
7. *ELASTICITY IN AIM, SIZE OF FLOCKS, AREA ETC.*
8. *PARENTS READILY COOPERATE.*

Fig. 1.—Chart showing value of poultry as a school subject.

(4) Management of the above without ownership.
(5) Management of the entire farm flock.

Ownership or partnership is very desirable, so that the interest may be keener and the lessons better learned.

Have an agreement (written, if possible) with the parents as to the plan of cooperation and the terms of ownership, profit sharing, etc.

---

[1] The term "home project" applied to instruction in elementary and secondary agriculture, includes each of the following requisites: (1) There must be a plan for work at home covering a season more or less extended; (2) it must be a part of the instruction in agriculture of the school; (3) there must be a problem more or less new to the pupil; (4) the parents and pupil should agree with the teacher upon the plan; (5) some competent person must supervise the home work; (6) detailed records of time, method, cost, and income must be honestly kept; and (7) a written report based on the record must be submitted to the teacher. This report may be in the form of a booklet. The club project should be identical with the home project from the school point of view.

Plan that each pupil who starts with scrub poultry shall try to work into pure-bred stock, if possible. This may be done by substitutions or at hatching time. Parents who are interested will vouch for the time records and accounts of the pupils. Arrange for the supervision of the work and, if possible, the occasional visit of some person expert in poultry raising. The local or county club leader or the county agricultural agent may help to arrange for this supervision.

POULTRY HOME PROJECT.

ADVANTAGES AT HOME:

1. INVESTMENT MAY BE MODERATE.
2. INCOME BEGINS EARLY.
3. LABOR SUITED TO YOUNG PERSONS.
4. UTILIZES HOME AND FARM WASTE
5. WEED SEEDS AND INSECTS DESTROYED.
6. CONTINUALLY SUPPLEMENTS FOOD SUPPLY.
7. POSSIBLE ON LIMITED AREAS.
8. SUITED TO BOTH BOYS AND GIRLS.
9. DISPOSAL OF ANIMALS NOT REQUIRED FOR PROFIT.

FIG. 2.—Chart to balance figure 1, from the home point of view.

Whenever projects are not taken, have as many practical exercises as possible to illustrate the lessons.

THE PROJECT REPORT.

Whenever any single phase of the project is completed, the pupil should be required to write a report on that section. A skillful teacher will be able to have some of these reports written as language exercises under such titles as "How I selected my poultry flock," or "Feeding my flock of laying hens." The labor, feed, egg, and other records should be compiled and balanced each month as arithmetic practice. Toward the end of the school year have the pupils assemble these reports and records, together with selected pictures, and fasten them within heavy covers as the "booklet" or the final project report. See forms pages 30–32.

ORGANIZED CLUB PROJECTS.

The State agent in charge of club work at the State agricultural college will cooperate with the local teacher in the organization of poultry clubs and will usually help to arrange for the supervision of home work, which is the most difficult problem in practical agricultural instruction. Farmers' Bulletin 562 is devoted to this subject. New poultry clubs are usually organized about January 1, but the pupils should not postpone all home work until that time.

School credit for the home project should preferably be given as a part of the rank for the course in poultry. In such a case the home project becomes fundamental and the school lessons a means toward accomplishing the practical end, so that the rank and credit are given on the work as a whole. The weight given to this credit is not to be in proportion to the number of fowls kept, but, assuming that enough are kept to make the work worth while, in proportion to the phases of poultry management developed and the application of principles learned. The subject of school credit for home practice in agriculture is developed in United States Department of Agriculture Bulletin 385.

CHARTS AND CHART MAKING.

Most of the illustrations in this bulletin are chosen with a view to showing teachers how to provide charts, models, and other illustrative material as well as how to illustrate poultry topics. Large sheets of manila paper or cloth charts may be used. Lettered charts, such as figures 1 and 2, may be used to emphasize certain principles. District surveys may be arranged on chart forms such as those given on page 25, but should also be put on a copy of the district map as is suggested in the census map in figure 3. Do similar work for other animals and crops.

Relative gains and losses, total productions, and other results may be charted by the graphic method shown in figure 10 (p. 17), especially where the fluctuations during a period of time are not taken into account. Fluctuations may best be shown on a vertical chart as in figure 7 (p. 13). Have the pupils practice first in charting the rise and fall of the daily temperature and then make graphic records of the laying of their home flocks. This will enable them to interpret such charts as figure 8 which shows the effect of animal feed in the ration.

Pictures selected to show contrasts may be cut out of farm papers and mounted as in figure 4. In some cases it may be desired to contrast two varieties of fowls which are somewhat alike.

Such charts as figure 5 may be enlarged with or without the lettering and used for class drill. Have pupils assist in this chart-making and apply the same ideas to other animals. Such a chart as figure 13 (p. 22) is not intended to take the place of practice in the process but may be used for a preliminary lesson, and, by remaining before the class for some time, will fix the facts firmly in mind. A convenient size is about 24 inches wide, the length being determined by the subject.

Such illustrations as figure 9 are best made with objects instead of pictures. Have the facts at hand to teach the whole lesson. Compare average poor results with results which are reasonably obtainable. In figure 9 (p. 15) the best production represents only a 50 per cent yield, which however is quite good and is the average of 2,000 hens. Such a comparison does not invite criticism.

Have pupils make working drawings of all equipment before making the article itself. Such drawings as figure 8 make good charts, and models should be made at school or at home. Models of small articles should be made full size, but it is often desirable to construct a model of a new type of house at school, leaving one portion unfinished as is shown in figure 6.

In every case the teacher should use the pupils so far as possible instead of doing all these things himself. Some very helpful charts may be copied from those issued by commercial concerns.

THE DISTRICT SURVEY.[1]

At the beginning of the school year the teacher should obtain all the information possible as to the poultry in the district. The pupils may assist in obtaining this information, but it is quite essential that the teacher become personally acquainted with his district. When the first data are collected two large charts should be prepared, one on a ruled form and one on a map of the district. These may be prepared and filled in by the pupils under the direction of the teacher. If no map of the district is to be found, one may be drawn and corrected by the pupils and several copies made on large sheets of manila or oak-tag paper for survey purposes. On one map indicate the homes of the pupils. Use small colored seals or pins to indicate the breeds or varieties of fowls on each farm, and mark the number kept. (See fig. 3.) Later indicate sources of feed, method of incubation, and other data.

Fill the first chart in September but collect further data as new practice becomes seasonal. Some of the following data should be obtained concerning each farm in the district: Size of farm, location, owner's name; breed, variety, and strain of poultry kept; pure-bred

---

[1] Suggested survey forms will be found on page 25.

or scrub; number of adult hens, adult males, this season's chickens at date; method of incubation; aim, commercial or domestic; number of fowls to be fattened for market; improvements planned for this season.

REFERENCES.

Nearly all the subject matter for class discussion and instructions needed for home-project work will be found in bulletins available either free or at a nominal cost. Nearly every State agricultural college has published one or more poultry bulletins, and in many cases the extension service of the college has issued circulars suited to school use. These should be obtained by addressing the dean of the agricultural college.

The Farmers' Bulletins of the United States Department of Agriculture cited in this bulletin cover most of the topics to be studied

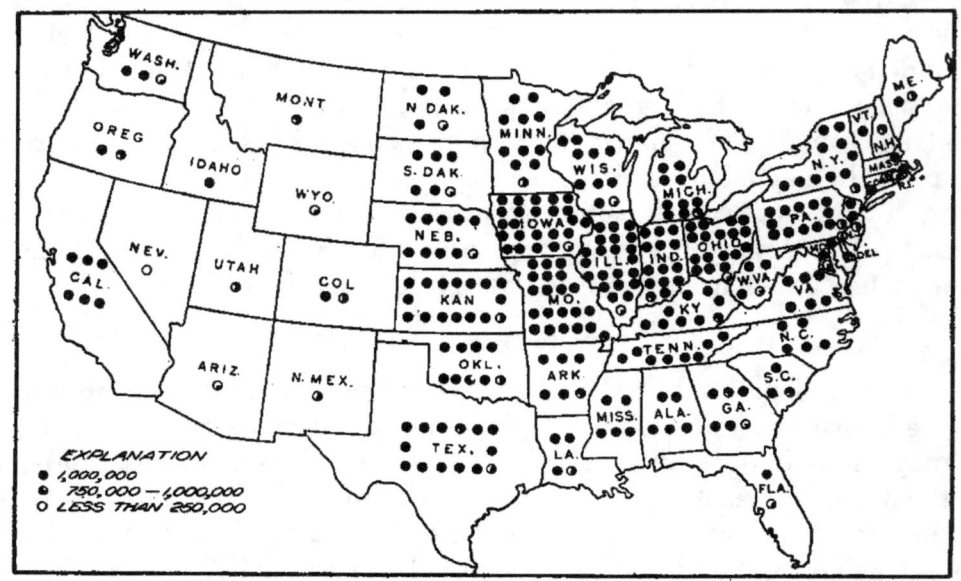

FIG. 3.— Map chart of all fowls on farms—census of 1910. (Similar charts of farms in district are suggested.)

and are suited to the pupils' reading and study. Bulletins in this list will be sent free, so long as the supply lasts, to any resident of the United States, on application to his Senator, Representative, or Delegate in Congress, or the editor and chief Division of Publications, United States Department of Agriculture, Washington, D. C. Because of the limited supply, applicants are urged to select only a few numbers, choosing those which are of special interest to them and ordering but one copy of each. When this free supply has been exhausted, a limited number are yet for sale. One should apply to the Superintendent of Documents, Government Printing Office,

Washington, D. C., who has these bulletins for sale at 5 cents each. Some other publications of this department are for sale by the Superintendent of Documents, but these are more often technical bulletins, of interest only to those who wish to specialize in the subject. Classified lists of department publications on different phases of agriculture for teachers' use are issued by the Agricultural Instruction Division, States Relations Service, United States Department of Agriculture.

Several of the textbooks on poultry raising are suited to the needs of the teacher, and some of the more elementary texts may be used by the pupils for reference.

Good pamphlets are issued by some of the poultry-supply houses, and some are supplied free to teachers. One of the best ways to keep these bulletins ready to use is to file them in pasteboard cases which may be made for the purpose, grouping the bulletins by subjects. Some poultry journals are usually to be found in the pupils' homes, and the poultry columns of other farm papers will be found helpful. Encourage pupils to bring these to school.

ILLUSTRATIVE MATERIAL.

On a chart of the size used for surveys have some pupil copy a diagram showing the points of a fowl. Collect pictures of prize-winning fowls, and diagrams and photographs of poultry houses and equipment. Procure and preserve samples of as many kinds of poultry feed as may be of local interest. The boys may be glad to make models of some needed poultry equipment. (See Farmers' Bulletins 586 and 606 for information about collecting and preserving illustrative material.)

Modifications due to climate in different sections must be made in considering any bulletins not prepared especially for the locality in which the school is located. Such modifications concern types of shelter, dates of hatching and brooding, green crops, and especially winter management. For this reason it is wise to depend as far as is convenient on the publications of the State agricultural college and experiment station and to consult the poultry experts of that institution. The State club leader will give practical assistance along these lines, especially in the matter of home practice.

Special phases of poultry raising for districts raising ducks, geese, turkeys, and guinea fowls are not developed in this bulletin. The method of study here outlined may be modified by the teacher to fit such cases, and Farmers' Bulletins 200, 684, 697, and others, with bulletins published in the State, will give the subject matter needed.

LESSON ONE.

SUBJECT: SELECTING THE FLOCK.

EARLY SEPTEMBER.

Topics for study.—(Develop only those which may have local application.)

(1) The individual fowl: Characteristics indicating strong constitutions, faults or defects to be avoided, value of desirable ancestry, relative value of hens and pullets. (See fig. 4.)

(2) The breed and variety: Develop the application of the terms *breed, class, variety,* and strain as applied to poultry. Find what varieties and strains are considered successful locally and study these more carefully. What color of egg does the local market demand?

FIG. 4.—Evidences of strong and weak constitutions, showing method of arranging pictures for contrast.

Consider the aims of the pupils in their projects in covering this topic.

References.—Farmers' Bulletins 51; 287, pp. 5, 6; 355, pp. 29–33; 562, p. 7. Also station and extension bulletins and circulars of the State agricultural college. "The American Standard of Perfection" is the authority for judging breeds and varieties.

The home project.—Have each pupil begin the project at once. Decide on aim in each case—eggs for market, meat for market, eggs for hatching, eggs and meat for home. Help pupils to obtain expert advice in selecting their project flocks. Cull at once and plan to dispose of the culls in accordance with suggestions of Lesson three.

Material and exercises.—Use photographs and other pictures of poultry. Typical feathers showing markings of breeds, etc., may be collected and mounted. Complete and use the survey charts and maps. The teacher should visit one of the fairs with his class, if possible, and study the poultry as well as other exhibits. Take a class excursion to a farm where a successful poultryman will assist the class.

Correlations.—Utilize the related topics in the other school subjects. Examples of this may include the history of the breeds and

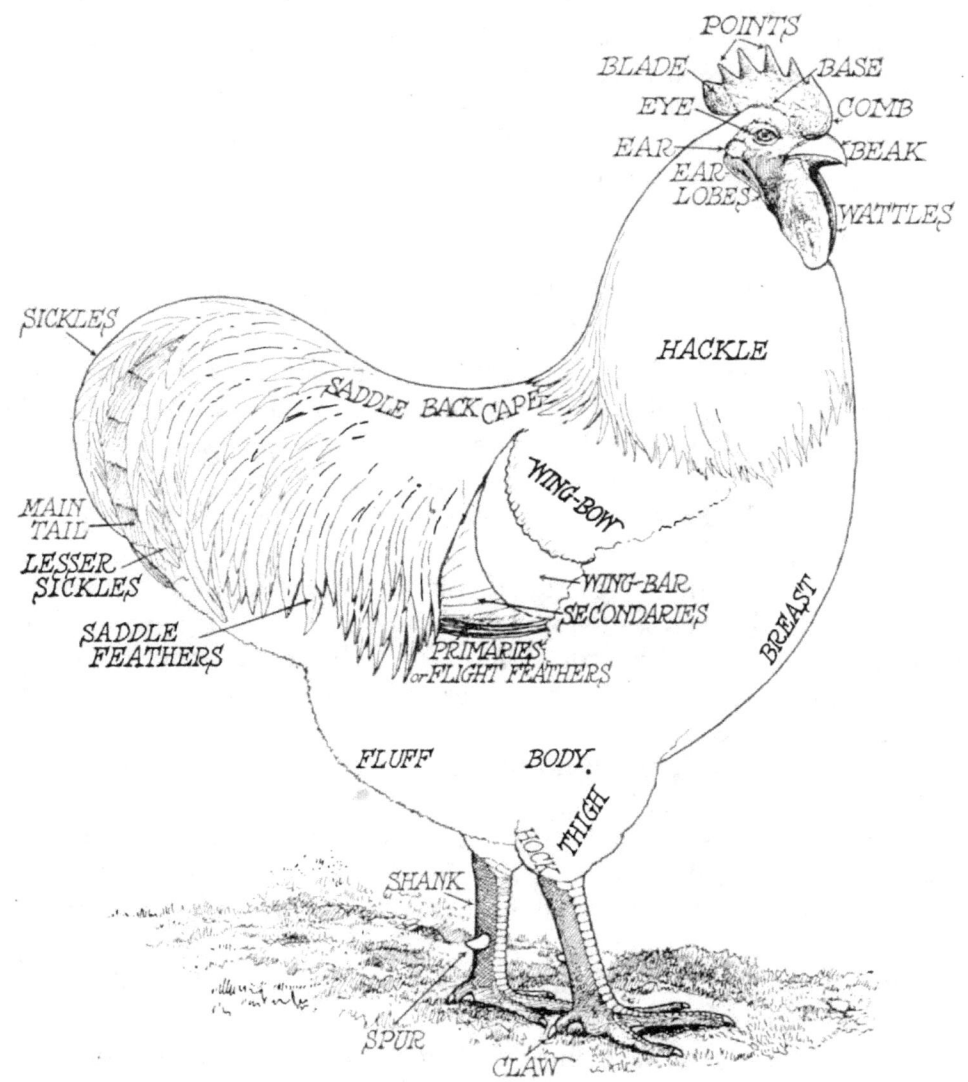

Fig. 5.—Glossary chart giving the names of the various sections of a male fowl. A study chart. For drill, omit lettering.

Note.—In the female the cushion takes the place of the saddle of the male and the sickle feathers are absent.

the geography of their origin. Language lessons may include descriptions of typical fowls or a report of observations at the county fair. In arithmetic consider the gain on 150 eggs yearly per hen when the eggs of the required color sell for 3 cents a dozen more than other eggs. Find how many more eggs per year will be required to offset extra price for pure-bred pullets. Compute number of extra eggs needed to offset the value of 3 pounds 4 ounces extra weight on each hen, poultry selling at the local rate. Have pupils enlarge figure 5 as a chart, possibly not putting the names of points on the chart.

Make out inventories for project or club work, including value of fowls, house, and equipment. (See also U. S. Department of Agriculture Buls. 132, Correlating Agriculture with the Public School Subjects in the Southern States, and 281, Correlating Agriculture with the Public School Subjects in the Northern States.)

LESSON TWO.

SUBJECT: DISPOSING OF CULLS.

EARLY FALL.

(Omit this lesson if no pupil in the class has an opportunity to apply it in his project or with his parents' flock.)

Topics for study.—What shall be culled out of the farm flock? What older fowls shall be kept? How many males are needed? Are the culls for market or for home use? Fatten and market as early as possible. Show loss from boarding the culls too long. What will the market accept? How does the local market demand that the poultry shall be prepared?

Fattening rations. Restricted range. Shipping alive. Dressed poultry. Individual customers. Parcel-post shipment. Local market prices. Poultry for home use. Value of poultry as food. The economy of a home-grown meat supply. Select topics needed and read references.

References.—Farmers' Bulletins 287, pp. 34–41; 355, pp. 29–39; 562, p. 11; 635, pp. 11, 12, 21.

Home projects.—Consult with each pupil and plan that each one may carry out such phases of this lesson as fit his case. The older pupils would gain much from the business practice involved in marketing.

Exercises.—A trip to observe marketing methods is helpful whenever pupils are to have this work to do.

Correlations.—Compare the price the farmer gets for poultry with the retail price. Compute the percentage loss to farmers who sell poultry and buy meat at retail. Weigh a flock of culls and compute present market value. Find present daily cost of food and compute gain or loss if the present flock gains 1 pound per fowl and sells at 1 cent more per pound on November 20. Demonstrate that it pays to sell early, but to keep a good supply of meat for the home. Compute each bill of poultry sold by pupils.

Have pupils write letters requesting such bulletins as each needs.

LESSON THREE.

SUBJECT: POULTRY HOUSES AND YARDS.

OCTOBER.

Topics for study.—Divide the study if some of the pupils are to construct new houses.

(1) Old houses: Drain the floor, replace earth and litter with fresh material. Procure sufficient light and ventilation. Clean out all rubbish. Disinfect roosts, nests, and other equipment. Whitewash the inside walls of the house. Be sure roof will not leak. Reconstruct any roosts, nests, etc., which are unsuitable.

(2) New houses: What location for light and drainage? What relation to yards and other buildings? How much floor space? Style of house. Best roof for the location. What arrangement of

FIG. 6.—A model of a poultry house used on the Government farm at Beltsville, Md., suggesting how pupils may make models.

doors, windows, and open spaces? Ventilation how? Best arrangement of nests, roosts, and equipment.

(3) Yards and fences: Value of free range. Value of alternate yard system, one growing crop and one being pastured. Relation of fencing to the breed of fowls. Fencing for or against chicks. Fall green feed in yards. Fall sowing in alternate yard for spring feed (see p. 28). Value of the orchard as a range.

References.—Farmers' Bulletins 574; 287, pp. 7-19; 355, pp. 21-25; 682; 480, pp. 8-16. Nearly every State has literature on houses adapted to its own climate.

Home projects.—Each pupil should arrange at once for the proper housing of his flock. Have each apply the lessons learned from the references. Keep daily accounts of expense, feed, labor, and income. See forms pages 30-32.

Material and exercises.—Select pictures and plans of houses from reliable sources. Have pupils inspect some good houses and make a class excursion for this purpose where it is feasible.

Correlations.—Descriptions of buildings and equipment or the story of making the house will provide good language work. Plans, perspectives, and working drawings are practical work. Problems in areas, lumber bills, and inventories should be drawn from the projects of the pupils.

Manual training should include making nests and feed hoppers. Models of good styles of houses may be made, especially the model of the State college type. Such a model developed in the United States Department of Agriculture is illustrated in figure 6. This type of house has been made and used under the direction of the Poultry Division of the Bureau of Animal Industry. It has been found satisfactory on the Government farm at Beltsville, Md. A model of a selected type may be made by the pupils.

LESSON FOUR.

SUBJECT: POULTRY FEEDING.

NOVEMBER.

Topics for study.—Elementary principles of feeding. Available nutrition in different material used for poultry foods. Requirements for growth, renewal of tissues (especially feathers), warmth and egg production. Balancing a ration. Substitutions to save expense. Sources of animal food.

The necessity of animal food is plainly shown in figure 7, in which the production of eggs is seen to be much greater in pens fed on either beef scrap, fish scrap, or skim milk than in the check pen where the ration contained no animal food.[1]

The rations used in obtaining these results over a period of four years were as given below:

Rations to test the effect of animal food for poultry.

Ration.	Check pens.	Skim-milk pens.	Meat-scrap pens.	Check pens.	Skim-milk pens.	Fish-scrap pens.
	Pounds.	Pounds.	Pounds.	Pounds.	Pounds.	Pounds.
Corn	10	10	10	10	10	10
Wheat	10	10	10	5	5	5
Oats	5	5	5	10	10	10
Bran	5	5	5	5	5	5
Shorts	5	5	5	5	5	5
Skim milk		62			50	
Meat scrap			3.5			
Fish scrap						3.6

[1] Adapted from Purdue University, Indiana, Bul. 182, Nov. 1915.

The material used in the rations was such as might be practical on the average Indiana farm. More ideal rations are often too expensive. To these rations were added grit, shell, and green food. The chart

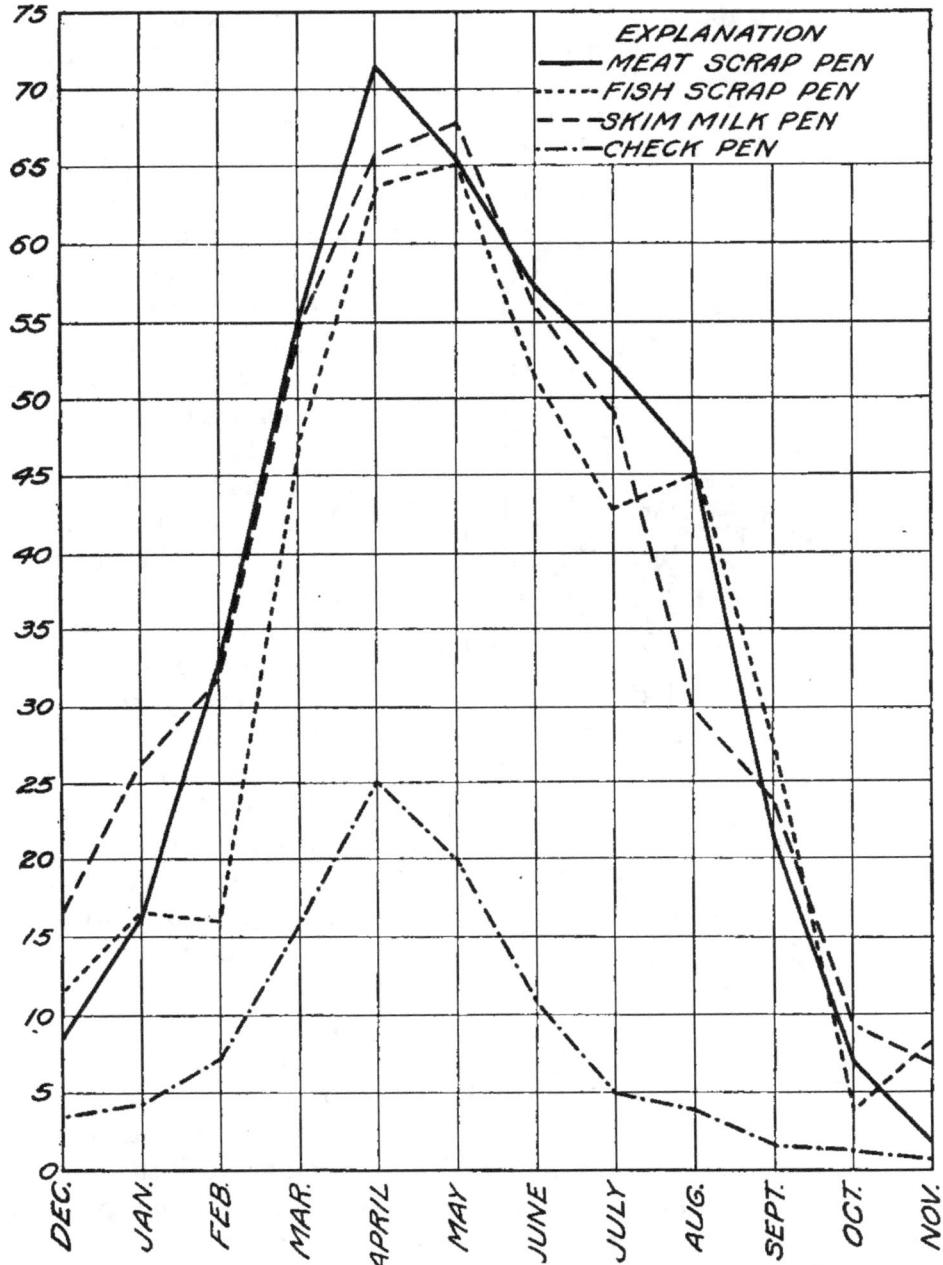

FIG. 7.—Average per cent egg production per month per pullet, with and without animal food. Graphic type of chart for results of several factors. (Adapted from investigations of Agricultural Experiment Station of Purdue University, Lafayette, Ind., Bul. 182.)

does not bring out the relative value of meat scrap, fish scrap, and skim milk, but shows the necessity for some animal food. Skim milk is often cheaper than other forms of animal food.

Value of dry mash, green food, grit, shells, charcoal. Why a scratch feed? Why have a variety? Quantity of feed needed? How obtain succulent food? How provide water? To what extent shall wet mashes be used? How feed the layers and the breeding flock? Avoid diseases of the digestive tract. Avoid spasmodic feeding of green foods which leads to overeating. Feed hopper approved by the U. S. Department of Agriculture is shown in figure 8.

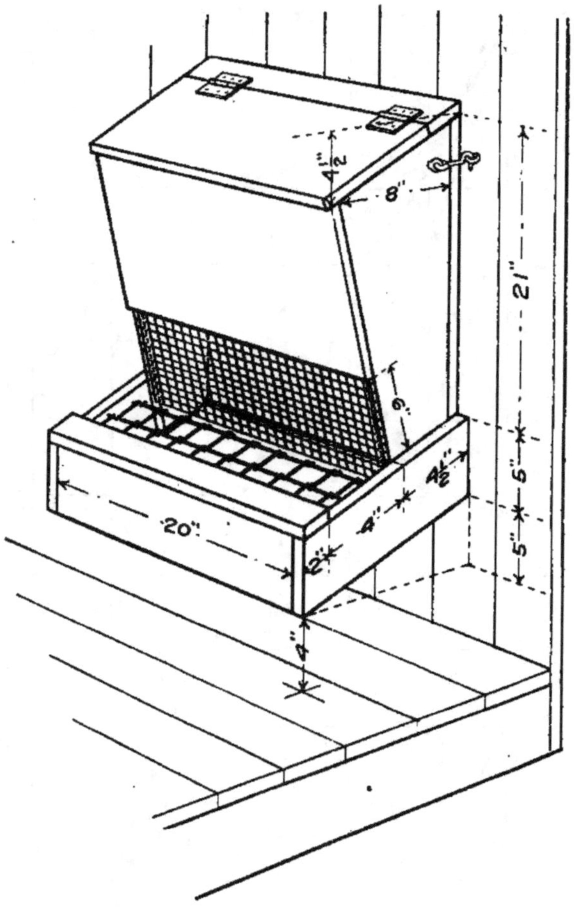

References.—Farmers' Bulletins 287, pp. 20–26; 355, pp. 35, 36; 528, p. 10. State agricultural college bulletins are especially valuable on feeding.

Home projects.—Have each pupil arrange for feeding his flock rations which best fit his own circumstances, considering the available home-grown feed, local prices of grain, supply of table scraps or other by-products to be obtained. The purpose of the flock should be kept in mind as well as the relative protection from the winter's cold. Have feed records kept constantly and reported occasionally. Weight should not be estimated. Labor record should be kept. See record forms, pages 26–32.

FIG. 8.—An efficient feed hopper, tested by the Bureau of Animal Industry. (Suitable drawing and manual correlations are suggested.)

Material and exercises.—Have samples of as many poultry foods as possible brought to school. Classify and study these and arrange them for the permanent school collection. Have some members of the class post each week the list of prices of all the grain and other feed for poultry, using local retail prices. Make a list of home-grown feeds and give the price at which the farmer might sell each. Change this list as prices change.

Correlations.—Compute the cost of different rations used. More mature pupils may compute the balanced rations, but so many mixtures have been computed by experts that it is not necessary for the amateurs to learn this detail.

Lessons on human hygiene, food, and digestion should be correlated with those on poultry feeding, since the same principles apply in each. A knowledge of one topic leads to an appreciation of the other. In drawing classes have plans made for equipment (as in fig. 8), and manual-training classes should make models to be used at home.

LESSON FIVE.

SUBJECT: WINTER MANAGEMENT OF POULTRY.

EARLY DECEMBER.

Topics for study.—Consider the need of dry floors, proper litter, dust boxes. Provide ventilation. Use of open front instead of windows.

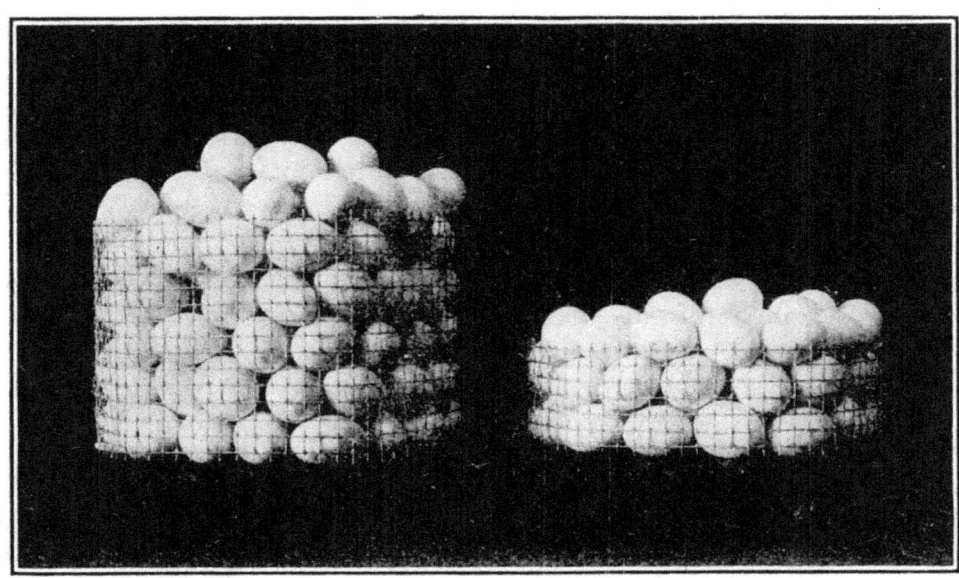

FIG. 9.—An exhibit arranged to contrast what the average farm hen lays and what selected hens lay when properly cared for. Ratio, 85 to 151.

Value of sunlight. Ample range, especially for breeding flock. Provide exercise for all. How control range of temperature? Cleaning nests, litter, and dropping board. (See management of poultry manure on p. 29.) Watch roosts and nests for mites, lice, and spiders. How disinfect? How sterilize water fountain? Variations in diet during very cold weather. Watch for signs of broodiness and use "broody coops." Separate breeding stock from other laying hens. Keep males from flock which lays market eggs. Keep a daily record of everything. The importance of records may be seen by referring to figure 9.

In five egg-laying contests held in 1915, 2,375 hens laid an average of 151 eggs, while it is estimated that the average for farm hens in this country is not over 85 eggs per year. This average of 151 eggs included several hens which laid no eggs and several which laid over 200 eggs, the highest record being 314. It seems fair to claim that the farm hens ought to average 151 eggs, as that is about 42 per cent of a perfect score and is less than 50 per cent of the highest record actually made. Without doubt both breeding and management figure in the improvement over average records.

A graphic presentation of the 151-egg yield compared with the 85 eggs which the average farm hen lays is suggested in figure 9.

References.—Farmers' Bulletins 287; 355, pp. 34–38.

Home projects.—Each pupil should carry out every point in management which applies to his project. Records of labor, feed, eggs, and sales should be kept with care and brought to school for the correlation work. If diseases or pests appear, consult the references and call on specialists at the State college of agriculture.

Exercises.—Have pupils individually visit the more prominent poultry establishments and report on the management. An evening talk or illustrated lecture for the public may be arranged. Farmers' institute lectures with lantern slides are available in the States Relations Service, United States Department of Agriculture.

Correlations.—Use the pupils' labor, feed, and income records for arithmetical problems. Have each pupil make a balanced statement of cost and income at the close of each month. Have the story of the management of the month written with care and credit it as a language lesson.

LESSON SIX.

SUBJECT: POULTRY DISEASES AND PESTS.

DECEMBER.

Topics for study.—(1) Digestive disorders: Prevent by using clean utensils, pure water, clean food, balanced ration, moderate use of new foods, keeping sparrows away.

Signs: Lack of vigor, droppings unusual in appearance or odor, failure to feed well, unnatural appetite.

(2) Diseases due to poor ventilation, overexposure, and dampness, may be prevented in most cases. Contagious diseases may appear and the fowls so affected should be isolated and expert advice gained. A discussion of different diseases is not necessary, but any disease which becomes prevalent should be studied.

The treatment of lice, mites, spiders, and other animal pests is found in the publications used for references. Banish the English sparrow, provide a dust box, keep things clean, and disinfect frequently.

References.—Farmers' Bulletins 530; 287, pp. 42–47; 528, p. 11; 493; Bureau of Entomology Circulars 92, 170.

Home projects.—Careful watching and preventive practices are important. Pupils should take particular notice of each bird each day. Clean roosts and nests, sterilized feed and water utensils, pure water and clean food will prevent many troubles if the pupils persist in such practice. Neglect to open windows during the day results in high temperature and poor ventilation, rendering the fowl susceptible to colds, roup, and other troubles. Help pupils to get advice as to any diseases which appear.

Correlations.—The school course in physiology and hygiene includes topics in relation to digestion, respiration, ventilation, and general sanitation. Many of these principles apply equally well to poultry and other animals, and this correlation will prove helpful to both subjects. If a poultry club is to be organized, have the pupils write letters applying for membership, information, etc.

LESSON SEVEN.

SUBJECT: EGG SELECTION.

JANUARY.

Topics for study.—(1) Separation of breeding flock from layers. Elimination of males from laying flock. (See fig. 10.) Frequent, regular, careful collection of eggs.

(2) Market eggs: What to select. Clean, uniform eggs. Test any doubtful eggs. Avoid freezing. Grade to conform to best

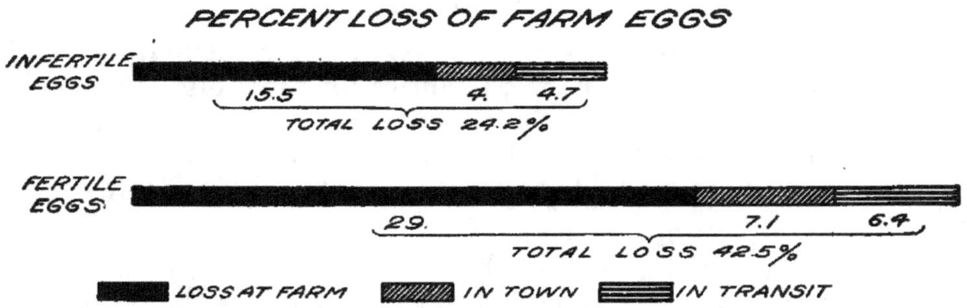

FIG. 10.—Horizontal graph teaching the desirability of infertile eggs for market.

local market. Show the loss resulting when a few low-grade eggs lower the grade of a whole case. How shall we dispose of the culled or defective eggs?

(3) Eggs for incubation: The source. Selection and breeding. Use only high-class fowls. Begin now to grade up to pure breds. Pure-bred males from other flocks are desirable. Necessity of vigorous constitutions. Handling the eggs. When begin to save for hatching? How long may they be kept? What temperature is best? It is much better to buy eggs for hatching than to use eggs from scrub fowls.

References.—Farmers' Bulletins 528; 562; 287, p. 40; 594, pp. 1–4; 682. The State agricultural college is the best authority concerning the climate and other local factors which may determine the time for hatching. Consult State agricultural college bulletins.

The home projects.—Have each pupil learn to apply methods adapted to his needs. Clean nests, frequent collections, and careful handling of eggs, sorting and packing for market are things a boy may grow careless about. Have all plans for incubation made before it is too late to do this well. A few well-selected fowls in a breeding flock may prove the nucleus of pure-bred flocks for the whole community. The lesson taught in figure 10 should be deeply impressed, because the margin of profit is small, usually due to the loss of fertile eggs. While this loss is greater in summer, yet the same principle holds at all seasons.

This is the month to organize or reorganize a poultry club. Refer to Farmers' Bulletin 562 and address the State leader of club work (at the State college of agriculture), requesting his assistance.

Exercises.—Demonstrate egg testing. Have pupils practice this first at school and then at home. Have them bring to school samples of eggs which have defects that make them unmarketable. If possible, have some successful poultry breeder give the class a talk on these topics.

LESSON EIGHT.

SUBJECT: INCUBATION.

EARLY FEBRUARY.

Topics for study.—What is the best time for incubation in this district, considering the climate, facilities for brooding, etc.? What influence has the market on the time for incubation? What should be known of the breeding stock? What eggs to discard? Care of eggs for hatching. Selecting the hen and preparing to set her. The process of setting the hen. How feed and care for the setting hen? Testing the eggs for fertility about the seventh day. Management during the last three days and bringing off the brood.

In case any of the pupils are to use an incubator, study its principles and manner of working. How start the incubator? Temperature control, testing the eggs, moisture control, final steps in hatching. How dispose of infertile eggs?

References.—Farmers' Bulletins 585; 562, pp. 8–10; 528, p. 8; 287, pp. 27–30; 682. Use State agricultural college bulletins constantly. Commercial houses, poultry-supply dealers, and others publish much useful material.

The home projects.—Have the pupils carry out with care the best obtainable advice on each step connected with this part of their projects. Incubation is a critical process. It is the opportunity to introduce pure-bred stock. Carelessness during the incubation period endangers the success of the next year of the project. This part of the project should rarely be omitted, although there are circumstances under which it is wise to buy chicks instead of running an incubator.

Exercises.—Utilize pamphlets issued by poultry-supply houses. If many pupils are to use an incubator, it may be desirable to have a demonstration at school or elsewhere.

The teacher should not take the authority to have an incubator run at school, because of fire risks.

Correlations.—Locate the sections which make a specialty of high-grade eggs for incubation, using advertisements and articles in poultry papers for information. Inquire into the history of artificial incubation. Compute the added profit and rate of gain for each pupil if all the eggs from his flock for one stated month had been sold for incubation at $1 a dozen.

LESSON NINE.

SUBJECT: MARKETING EGGS.

EARLY MARCH.

Topics for study.—What does the market demand as to color, grade, and style of packing of eggs? How gather? How sort and grade?

FIG. 11.—Some of the apparatus needed in teaching parcel-post shipment of eggs.

How and when test eggs? What shall be done with eggs from a stolen nest? Be sure to remove males from laying flock.

Delivery of eggs. Individual customers. Selling to local retailers or to hucksters. Dealing with commission houses. The community egg circle. Stamping and dating eggs. Shipping by parcel post; by express. Protection from heat and from other damage. Avoid loss from poor grading and obtain a reputation for a high-class article. Sell frequently, especially on a falling market.

References.—Farmers' Bulletins 287, p. 40; 528, pp. 10, 11; 562, pp. 10, 11; 594; 656; 682; Farmers' Institute Lecture No. 17, U. S. Department of Agriculture.

The home projects.—Have the pupils who are to market their eggs learn and put into practice the principles laid down in the references and make note of any gain resulting from correct practice. This practice of the pupils may initiate better practice throughout the community. A community reputation may be a valuable asset.

Practical exercises.—Procure and examine samples of shipping crates, especially for small lots. Keep this as a school exhibit as long as is wise. Collect and tabulate, as a part of the survey, information as to egg production and marketing in the district.

Correlations.—Have pupils write out as a language exercise a description of the methods used in marketing eggs. Have problems

FIG. 12.—Brooder practice on the Government farm at Beltsville, Md. Observation of such practice is desirable for each class.

based on the project records of the pupils to demonstrate the value of the methods advised by the expert poultry men. Figure 11 illustrates facts in parcel-post shipments, although dozen lots do not make economical shipments. A general discussion of parcel-post shipments is found in Farmers' Bulletin 703.

On the State and county maps locate the markets for local eggs and the routes used in reaching them. Use inks of different colors.

LESSON TEN.

SUBJECT: BROODING OF CHICKENS.

MARCH OR APRIL.

Topics for study.—Take up the subject of natural or artificial brooding in accordance with the projects of the pupils and the practice of the district. How care for the chicks the first 48 hours after hatching? When, what, and how shall they be fed? Early practice with hens; with incubator chicks. The use of hovers. Management of artificial brooders. How keep the brood free from vermin? Brood coops and management. (See fig. 12.) How feed the growing chickens? What range is desirable? Protection from birds and animals of prey. How avoid chicken diseases? (The date of incubation varies with the climate and the market demands, and this lesson should be taken up before the time for chicks to appear.)

References.—Farmers' Bulletins 287, pp. 30–33; 528, p. 8; 624.

The home projects.—Have individual conferences with the pupils and help each one to obtain the assistance he needs with his project. This stage of the work varies much and is critical. Take a field trip which shall include a visit to farms where good brooders may be seen. Take photographs of good brooders.

Call upon the county agent and the club leaders for help on the practical problems of the pupils.

LESSON ELEVEN.

SUBJECT: PRESERVING EGGS.

WHEN EGG PRICES FALL.

Topics for study.—When does it pay to preserve eggs? Methods used: Bran, lime, salt, limewater, water glass, cold storage. Water-glass method simple and effective. How much of a range between preserving and using price is necessary to warrant the preserving of eggs? Why should preserved eggs not be sold to merchants? May they be sold to individual customers, if facts are told? Read with care the method of preserving eggs. What cautions as to use of eggs preserved in water glass? Eggs preserved during the low-price period at less than 20 cents a dozen may be used by the family when the market price is high, thus releasing for sale the entire output of fresh eggs. When the market price is extremely high, a good profit may be made by selling some of the preserved eggs, properly designated, at a price somewhat below the price of fresh eggs.

References.—Farmers' Bulletin 287, pp. 41, 42.

The home projects.—After the method has been demonstrated in a school exercise have the pupils obtain the material and put down in water glass at least a few eggs. When this practice is once begun, the family may easily continue it. The beginning is the chief difficulty. It may be more economical for several pupils to buy the commercial water glass in larger quantities and divide it. One dealer sold pints for 15 cents, quarts for 25 cents, and gallons for 80 cents.

Practical exercises.—At school boil 1 quart of water to sterilize it. Mix with this when cool about 4 ounces of commercial water glass.

Fill a large-mouthed 2-quart jar with selected eggs and pour the liquid over them to fill the jar. Fill with a little sterilized water if necessary. Seal and label properly. After this has remained at the school long enough to make its impression, allow the pupil furnishing the eggs to take them home. Make a chart by copying figure 13 on large manila paper.

Correlations.—Make arithmetical problems out of the following data, substituting local prices if it is desired. April eggs average 20 cents a dozen, May eggs 18, June and July eggs 16, August eggs 20. October sales quoted 25 cents, November 30, December 35, January 40, February 35. One quart of water glass costing about 25 cents will make 10 quarts when diluted and will preserve as many as 12 or 15 dozen of eggs. Find saving if May eggs preserved as suggested are used in December. Preserve 10 dozen eggs in May, 20 dozen in

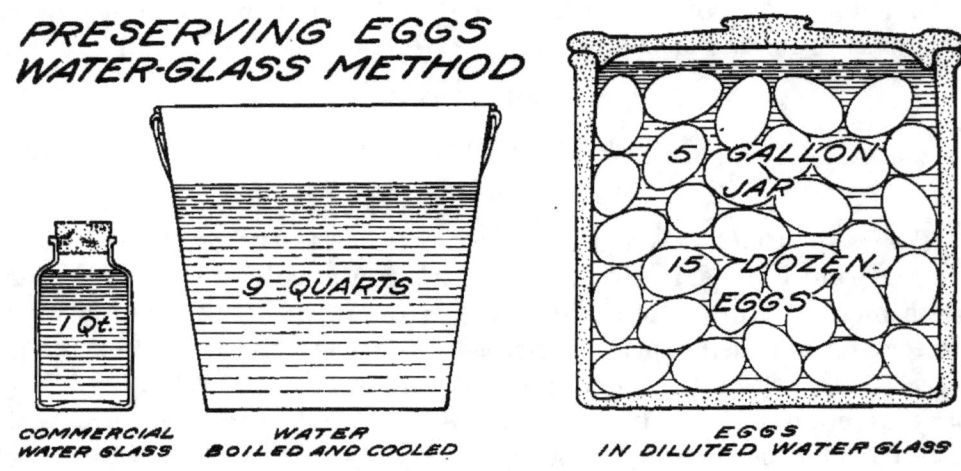

FIG. 13.—Chart to be used to fix in minds of pupils a method of preserving eggs.

June, and 10 dozen in July. Use half in January, half in February. How much is saved? When fresh eggs sell at 35 cents, preserved eggs will bring from 25 to 30 cents.

LESSON TWELVE.

SUBJECT: RAISING CROPS FOR POULTRY.

APRIL.

Topics for study.—(1) Crop rotations for forage in alternate yards. (See Supplement.) Oats, rye, alfalfa, clover, lettuce, Swiss chard, dwarf Essex rape, and other green crops may be sown in one yard while the flock uses the other. (See p. 28).

(2) Field and garden crops to be fed to poultry. Buckwheat as a crop for the orchard furnishes a useful poultry food. Corn and other grains, sunflowers, and alfalfa hay are among the possible crops for winter. Some succulent crops are especially needed for the

winter, and among those which may be raised are cabbages and mangel beets. Potatoes, turnips, and carrots are suitable food but are used in a more restricted way, usually the unmarketable culls.

References.—Farmers' Bulletins 22, 164, 278, 339, 424, 433, 485, and 537.

Home projects.—With two or more yards to be used in rotation have pupils raise crops for forage for the hens giving the hens the range of each yard in turn as the crops reach the proper stage.

Whenever it is possible, have the pupils plan to cultivate crops in fields or garden to provide feed for the poultry, especially succulent foods for winter use.

Correlations.—Have pupils draw maps or ground plans of their poultry yards and buildings. Have them study how to arrange for alternate yards or some system of furnishing green food. In arithmetic have them compute the average cost of raising crops for poultry and decide which they can afford to raise in view of the local prices paid for feed.

It is profitable for a pupil to take up some study and projects in plant production which may be correlated with the animal project. Corn clubs or alfalfa-club work may be suitable.

LESSON THIRTEEN.

SUBJECT: SUMMER MANAGEMENT OF POULTRY.

MAY OR JUNE.

Topics for study.—(Omit such topics as have no local bearing.) Summer range. Shade and shelter. Constant supply of green food. Infertile eggs after hatching season. Keep chickens growing fast. Dispose of surplus cockerels as early as possible. Look up market for broilers. Continue preserving eggs while price remains low. Cleanse and sterilize drinking fountains frequently. Watch for evidence of vermin and diseases.

References.—Farmers' Bulletins 287, 355, 528, 530, 562, 594, 624, and 682. State agricultural college bulletins and club literature, also textbook chapters on summer management should be used and should be available to the pupils during the summer.

Home projects.—The teacher should have personal conferences with the pupils on the matter of summer poultry management.

Have each pupil plan to carry out in the vacation months those practices in relation to feeding, care, marketing, etc., which best fit his circumstances. Summer supervision is a serious problem. Club leaders in cooperation with the State agent in charge of such work often solve this problem. A teacher retained for summer months supervises such work in some communities. Committees of parent-teacher associations or granges often assist. Wherever parents will agree to supervise the summer work, they are in a position to do it thoroughly.

Correlations.—Each pupil should complete as far as possible the report of his poultry project before the school term closes and should state his plan for the summer. This report may be made a language lesson. The accounts for cost of feed, labor, egg sales, etc., should be closed up to date and profit computed. These problems may well be taken up in the arithmetic classes. Have pupils write requests for such bulletins as they need at home in connection with their projects. Add to the survey charts any information obtained as to spring and summer practice, also any improvements.

SUPPLEMENT.

A DISTRICT POULTRY SURVEY.

[Sample forms.]

.... District.   Date—September, 1916.

Name of farmer.	Variety of poultry kept.	Total number.	Mature hens.	Mature males.	Intended for—		To winter—		
					Market.	Table.	Old.	Pullets.	Males.
C. Adams......	Rhode Island Reds	125	25	3	16	15	20	70	4
R. Davis......	Scrubs............								
Etc............									
Total......									

POULTRY CENSUS.

Date—September, 1916.

	Breed.	Number.	Breed.	Number.	Breed.	Number.	Total.
Egg breeds.....	White Leghorns.	275	Brown Leghorns	60	Black Minorca..	25	360
Meat breeds....	Light Brahmas..	74					
Dual breeds....	Barred Rocks...	183					
Total.....							

POULTRY HOUSE AND MANAGEMENT SURVEY.

Date—October, 1916.

Name of farmer.	Type of house.	Style of roof.	Style of front.	Fowls per pen.	Number housed.	Area per hen.	Type of floor.	Litter.	Nests.	Yards.
C. Adams...	2-colony.	Shed..	Open..	16–25	125	6 feet...	Gravel.	Straw.		Alternate.
R. Davis....										
Etc........										
Total .										

POULTRY-FEEDING PRACTICE.

[Survey form.]

Date—December, 1916.

Name of farmer.	Dry mash used.	Scratch feed.	Green feed.	Other feed.
C. Adams, Hill Road...	Bran, 60 pounds.. Corn meal, 30 pounds.[1] Meat scraps, 9 pounds. Salt, 1 pound......	Oats, 1 bushel [1]... Wheat, 1 bushel... Cracked corn, 1 bushel.[1] Buckwheat, ½ bushel.[1]	Alfalfa hay [1]........ Mangels [1].......... Cabbages [1]........ Sprouted oats [1]....	Sunflower seeds.[1] Oyster shells. Grit. Charcoal.
R. Davis, Maple Farm.				

[1] Home grown.

CHICKEN RAISING.

[Survey form.]

Date—February to May, 1916.

Name of farmer.	Source of eggs.	Date of setting.	Method of incubation.	Total hatched.	Method of brooding.	Comment.
C. Adams...	¾ home, ¼ bought.	Feb. 10 to Mar. 20.	Two 200 egg water incubators; 5 hens.	170	Acme brooders and hens.	Raises some broilers.
....	....	....	....	....	....	....
....	....	....	....	....	....	....
....	....	....	....	....	....	....

PUPIL'S MONTHLY SUMMARY OF LAYING FLOCK.

Total No. eggs for month, ...
Eggs sold.......... Prices........... Total............................$......
Eggs sold for setting..........at.......... Total............................ ......
Eggs used at home.......... Valued at............................ ......
Fowls sold.. ......
Fowls eaten at home.......... Valued at............................ ......

    Total income for month.. ......
Average number of hens........................
Average eggs per hen........................
Average eggs per day........................
Average income per hen........................
Cost of purchased feed used in month..$......
Value of home-grown feed used.. ......
Estimate of kitchen scraps (not charged)..$......

    Total value of feed for month.. ......
Gross profit of income over feed.. ......
Labor..........hours at.......... Total............................ ......
Other charges or losses, from daybook............................ ......

    Total charges.. ......
Net profit over all charges.. ......
Feed per hen, $.......... Total cost per hen............................ ......
Gross profit per hen, $.......... Net profit per hen............................ ......

A SPECIMEN MONTHLY ACCOUNT.[1]

Dr.

Date.	Item.	Feed.	Equipment.	Labor.	Miscellaneous.	Total.
1910.						
Apr. 1	100 pounds beef scrap..............	$3.00				$3.00
5	4 bushels shelled corn..............	3.00				3.00
5	5 bushels oats..............	3.25				3.25
8	Carpenter work on brooder..............			$2.00		2.00
8	Lumber..............				$4.00	4.00
10	1 indoor brooder..............		$8.00			8.00
10	6 gallons kerosene..............				.72	.72
19	300 eggs for hatching..............				5.75	5.75
20	100 pounds oyster shell..............	.80				.80
25	3 bushels wheat..............	3.75				3.75
	Total..............	13.80	8.00	2.00	10.47	34.27

[1] From Bureau of Animal Industry Circ. 176, A System of Poultry Accounting, p. 4

A SPECIMEN MONTHLY ACCOUNT—Continued.

CR.

Date.	Item.	Market eggs.	Hatching eggs.	Market poultry.	Breeding stock.	Total.
1910.						
Apr. 4	10 dozen eggs, at 24 cents	$2.40				$2.40
10	1 pen breeding fowls				$5.00	5.00
12	18 dozen eggs, at 24 cents	4.32				4.32
13	1 hen (home use)			$0.55		.55
14	3 market hens			1.80		1.80
15	8 dozen eggs (home use), at 24 cents	1.92				1.92
17	15 dozen eggs, at 23 cents	3.45				3.45
19	7 dozen eggs (home use), at 23 cents	1.61				1.61
19	300 eggs, hatching (home use)		$5.75			5.75
20	50 day-old chicks				3.00	3.00
23	7 dozen eggs (home use), at 23 cents	1.61				1.61
29	100 eggs, hatching		3.00			3.00
30	6 dozen eggs, at 22 cents	1.32				1.32
	Total	16.63	8.75	2.35	8.00	35.73

A BALANCE SHEET.[1]

[Yearly or more frequently.]

	Debit.	Credit.	Balance.
Value of inventory January 1, 1910	$347.30		
Interest at 6 per cent on capital invested, as represented by value of inventory above	20.84		
Expenditures during 1910	273.70		
Value of inventory January 1, 1911		$431.40	
Receipts during 1910		368.77	
Total	641.84	800.17	
Balance (profit)			$158.33

[1] From Bureau of Animal Industry Circ. 176, p. 6

AN ALTERNATE-YARD CROPPING PLAN.

One of the suggested crops is to be grown in one yard while a crop is pastured in the alternate yard.

	Yard 1.	Single flock house.	Yard 2.
April to July.	*Growing.* Oats. Chard or lettuce. Clover or vetch. Sunflowers (shade and seed). Cow peas. Rape.		*Feeding.* Winter rye. Winter vetch. Crimson clover (New Jersey and South). Sweet clover.
July 1 to October 1.	*Feeding.* Oats. Chard and lettuce. Clover or vetch. Cow peas. Rape.		*Growing.* Buckwheat. Dwarf Essex rape. Flat turnips.
October 1 to April 1.	*Growing.* Oats. Winter rye. Winter vetch. Sweet clover. Crimson clover.		*Feeding.* Buckwheat. Dwarf Essex rape. Flat turnips. Soy beans.

Select crops which will grow well in the given district.

Thickly sown crops for succulent food and summer shade. Adapted to climate of medium latitudes. Dates must be modified for extreme north or south. Consult local extension agents.

FRONT AND BACK ALTERNATE YARD PLAN.

1. Back yard. Permanent.
    Blue grass and clover or blue grass and alfalfa in sod.
    To be used as range while crops are growing in the front yard. Large area desirable.

Pen in continuous style house.

2. Front yard. Temporary crops.
    November 1 to April 1. Feed winter rye, vetch, crimson clover, etc.
    April 1 to July 1. Grow rape, chard, lettuce, buckwheat.
    July and August. Feed.
    September 1 to November 1. Grow winter rye, vetch, crimson clover, etc.

A ROTATION FOR POULTRY YARDS WHICH HAS PROVED PRACTICAL IN SOME LOCATIONS.

Date.	Yard A.	Yard B.
Mar. 1 to Apr. 30	Peas and oats	Feeding.
Apr. 30 to May 25	Feeding	Peas and barley.
May 25 to June 15	Dwarf Essex rape	Feeding.
June 15 to July 10	Feeding	Buckwheat and oats.
July 10 to Aug. 1	Buckwheat	Feeding.
Aug. 1 to Aug. 20	Feeding	Cow peas and millet.
Aug. 20 to Sept. 20	Rye, vetch, clover	Feeding.
Sept. 20 to Dec. 1	Feeding	Rye and vetch.

Special care must be taken lest the fowls return to the yard to which they have become accustomed.

GRAZING CROPS FOR POULTRY.[1]

[Adapted to the latitude of the southern boundary of Pennsylvania.]

Crop.	When sown.	Seed per acre.	Grazing period. Stage.	Duration.
Peas and oats	About Apr. 15	1 bushel peas, 2 bushels oats.	About May 20	Until full grown.
Chard	May 10 to July 1	3 pounds	8 inches–10 inches high.	Until consumed.
Rape	Beginning Apr. 20	6 pounds	6 inches–8 inches high.	Do.
Red clover	Aug. 20	12 pounds	About May 15	Until fed down closely.
Turnips	....do	3 pounds	Sept. 20	Until snow falls.
Buckwheat	May 10 to July 1	1 bushel	6 weeks	Until mature.
Soy beans	May 10 to June 10	....do	12 inches–15 inches high.	
Rye[2] and crimson clover.	Sept. 1	1 bushel rye, 15 pounds clover.	Graze early winter and spring.	
Sweet clover	Aug. 15 to Sept. 1	25 pounds	8 inches–10 inches high.	Until fed down or too tough.
Alfalfa	August	20 pounds	....do	Alternate periods.

[1] Suggestions by the Division of Forage Crop Investigations of the Bureau of Plant Industry. This phase of the investigations has received little attention from the viewpoint of poultry grazing, especially as to the relation of number of fowls to area of crops.
[2] Winter wheat may be substituted for rye. Farther north substitute hairy vetch for crimson clover.

MANAGEMENT OF POULTRY MANURE.

The manure produced is a valuable by-product of poultry raising. It is estimated that the average night droppings of a hen amount to 30 to 40 pounds per year. This represents the manure which can certainly be saved with the exercise of a little care. A conservative estimate indicates that this manure contains fertilizing constituents which would cost 20 to 25 cents if bought in the form of commercial fertilizers at ordinary prices. A flock of 100 hens would at this rate produce manure worth $20 to $25 per year. If, however, the manure is not properly cared for, as much as one-half of its fertilizing value is likely to be lost. To prevent loss frequent cleaning of the dropping boards is necessary, and some sort of absorbent should be used daily. The use in moderate quantities of fine, dry loam or road dust, or, preferably, mixtures of these with such materials as land plaster, acid phosphate, and potash salts has been recommended. Sawdust has also been used with good results at the rate of 10 pounds per hen per year mixed with 16 pounds of acid phosphate and 8 pounds of kainit. This gives a fertilizer which contains about 0.25 per cent of nitrogen, 4.5 per cent of phosphoric acid, and 2 per cent of potash, and is worth about $10 per ton at ordinary prices of these fertilizing constituents. It is a better balanced fertilizer than manure alone and is usually in better mechanical condition for application to the soil by means of fertilizer distributors or manure spreaders.

With the present high price of potash salts it is impracticable to use such materials in the way suggested, and it may also be impracticable to use acid phosphate. In this case somewhat larger amounts of sawdust should be used.

Sifted coal ashes may be used as an absorbent, but wood ashes or lime should not be mixed with the manure, as they are likely to cause the loss of its most valuable fertilizing constituent, namely, nitrogen (ammonia). Occasionally the litter from the poultry house may be mixed with the manure. This increases the bulk, but greatly reduces the value per pound of the manure and makes it difficult to apply to the soil, except where it is to be broadcasted and plowed in.

Poultry manure is more valuable than the manure of any other common farm animal, as the following table shows:

Analyses and value per ton of manure of different animals.

Animal.	Nitrogen.	Phosphoric acid.	Potash.	Value per ton.
	Per cent.	*Per cent.*	*Per cent.*	
Poultry	0.80 to 2.000	0.50 to 2.000	0.80 to 0.900	$7.07
Sheep	.768	.391	.591	3.30
Hogs	.840	.390	.320	3.29
Horses	.490	.260	.480	2.21
Cattle	.426	.290	.440	3.02

Poultry manure is particularly well adapted to gardening, and the pupils should either use it on their own garden projects or dispose of it at a good price, thus increasing the profits of their flocks.

POULTRY-CLUB REPORT FORMS.

[Furnished by the Bureau of Animal Industry, U. S. Department of Agriculture.]

These or similar forms are used by the club leaders of the different States and may be obtained through the State club leader of boys' and girls' clubs at the State college of agriculture.

REPORT OF HATCHING CHICKS.

Name.. Age......................
Post office................ County............... State......................
Name of poultry club................................... Year......................

| Lot. | Number of eggs. | Kind of eggs. | Date set. | Hen or incubator. | Eggs tested out. | | | | Eggs remaining. | Chicks hatched. |
					On seventh day.		On fourteenth day.			
					Infertile.	Dead.	Infertile.	Dead.		
1										
2										
Etc										

REMARKS.—If an incubator is used, state make of same and total amount of the kerosene consumed. Make of incubator,........................; kerosene used, ... gallons.

State the total number of hours of labor spent in attending hens or incubator: .... hours.

Two copies of this report must be filled out, one to be given to the teacher at the end of the weaning period of the last lot of chicks and the other to be retained by the club member. (This is to be done for all reports at the time they are filled out.)

REPORT OF BROODING CHICKS.

Name.. Age.......................
Post office.................. County.............. State....................
Name of poultry club.............................. Year......................

Lot.	Number of chicks.	Kind of chicks.	Date hatched.	With hen or brooder.	Date chicks were weaned.	Number of chicks died.	Number of chicks remaining.	Weight of lot when weaned.	
								Pounds.	Ounces.
1.....									
2.....									
Etc.....									

REMARKS.—If a brooder is used, state make of same (or homemade) and total amount of oil consumed. Make of brooder..
Kerosene used .... gallons. State the total number of hours spent in attending hens and chicks or brooder .... hours. State the kinds of feed and value of each.

Kind of feed.	Number of pounds used.	Cost of feed.	
		Dollars.	Cents.
....			
....			
....			

REARING REPORT FROM WEANING TIME UNTIL JANUARY 1.

Name.. Age.......................
Post office.................. County.............. State....................
Name of poultry club.............................. Year......................

Number of chicks at weaning time.	Kind of chicks	Chickens sold for market.		Total amount received.	Pullets raised to Jan. 1.	Cockerels raised to Jan. 1.	Number of chicks died.	Number of chicks remaining.	Total weight Jan. 1.
		Number.	Pounds.						Pounds.
....									
....									
....									

REMARKS.—Chickens sold for breeders: Cockerels—Number ...., value $......; pullets—Number ...., value $.......
Highest price received for a single bird: $......; cockerel or pullet ....
Exhibited at ...... fair. Number of birds exhibited ....; total prize money received, $.......
Number of special premiums ....; nature of same ....
Chickens used at home should be credited at market prices.
Labor at 10 cents per hour: Number of hours ....; cost $.......

Kind of feed.	Number of pounds used.	Cost of feed.	
		Dollars.	Cents.
....	....	....	....
....	....	....	....
....	....	....	....

YEARLY REPORT.

Name.................................... Age..................
Post office............... County.................. State..................
Name of poultry club................................. Year..................

Dr.

Month.	Cost of feed.	Cost of equipment.	Labor.	Cost of eggs for hatching.	Cost of breeding stock.	Number of hens died.	Miscellaneous.
January.........							
February........							
Etc.............							

Cr.

Month.	Market eggs sold.		Hatching eggs sold.		Market poultry sold.		Breeding stock sold, value.	Miscellaneous.
	Dozen.	Value.	Sittings.	Value.	Pounds.	Value.		
January.........								
February........								
Etc.............								

INVENTORY REPORT.

Name.................................... Age..................
Post office............... County.................. State..................
Name of poultry club................................. Year..................

Inventory January 1.

Stock: Number of cocks ...., value, $......; number of hens ......, value, $......; number of pullets, ..., value, $......; number of cockerels, ..., value, $....... Total value of stock, $........

Equipment: Value of poultry houses, $......; value of feed hoppers, $......; value of drinking vessels, $......; value of brood coops, $......; value of poultry fences, $.......; number of incubators, ...; make, ...; value, $......; number of brooders, ..., make ......, value, $......; value of grain on hand, $....... Total value of equipment, $.......

LIST OF PUBLICATIONS RELATED TO THE SUBJECT.

FARMERS' BULLETINS AVAILABLE AT THE DATE OF THIS PUBLICATION.

- 22. The Feeding of Farm Animals.
- 51. Standard Varieties of Chickens.
- 142. Principles of Nutrition.
- 182. Poultry as Food.
- 287. Poultry Management.
- 355. A Successful Poultry and Dairy Farm.
- 445. Marketing Eggs Through the Creamery.
- 452. Capons and Caponizing.
- 480. Practical Methods of Disinfecting Stables.
- 493. The English Sparrow as a Pest.
- 511. Farm Bookkeeping.
- 528. Hints to Poultry Raisers.
- 530. Important Poultry Diseases.
- 562. The Organization of Boys' and Girls' Poultry Clubs.
- 574. Poultry House Construction.
- 585. Natural and Artificial Incubation of Hens' Eggs.
- 594. Shipping Eggs by Parcel Post.
- 624. Natural and Artificial Brooding of Chickens.
- 635. What the Farm Contributes Directly to the Farmer's Living.
- 656. The Community Egg Circle.
- 682. A Simple Trap Nest for Poultry.

SPECIAL LINES OF FOWLS.

- 200. Turkeys.
- 234. The Guinea Fowl.[1]
- 684. Squab Raising.
- 697. Duck Raising.

SCHOOL METHODS.

- 586. Collection and Preservation of Plant Material for Use in the Study of Agriculture.
- 606. Collection and Preservation of Insects and Other Material for Use in the Study of Agriculture.
- Department of Agriculture Bulletin No. 132, Correlating Agriculture With the Public School Subjects in the Southern States.[1]
- Department of Agriculture Bulletin No. 281, Correlating Agriculture With the Public School Subjects in the Northern States.
- Department of Agriculture Bulletin No. 385, School Credit for Home Practice in Agriculture.

FARMERS' BULLETINS ON FEEDING AND CROPS FOR FEEDING.

- 22. The Feeding of Farm Animals.[1]
- 142. Principles of Nutrition and the Nutritive Value of Food.[1]
- 164. Rape as a Forage Crop.[1]
- 278. Leguminous Crops.[1]
- 318. Cowpeas.
- 339. Alfalfa.

---

[1] Available at 5 cents a copy; sold by the Superintendent of Documents.

424. Oats: Growing the Crop.
433. Cabbage.
443. Barley: Growing the Crop.
485. Sweet Clover.
537. How to Grow an Acre of Corn.
552. Kafir as a Grain Crop.
617. School Lessons on Corn.
686. Uses of Grain Sorghums.

The Yearbook of the Department of Agriculture has articles of value, some of which are published as separates and are sold by the Superintendent of Documents.

The Prevention and Treatment of Diseases in Poultry, 1911 Yearbook, pp. 177–192; also issued as Separate No. 559; price, 5 cents.

The Handling and Marketing of Eggs, 1911 Yearbook, pp. 467, 468; as Separate No. 584; price, 5 cents. This is an abstract of Bureau of Animal Industry Bul. No. 141, The Improvement of the Farm Egg.

Improving the Quality of Poultry and Eggs, 1912 Yearbook, pp. 345–352; as Separate No. 596; price, 5 cents.

Handling Dressed Poultry a Thousand Miles from Market, 1912 Yearbook, pp. 285–292; as Separate No. 591; price, 15 cents.

Effect of the Present Method of Handling Eggs, 1910 Yearbook, pp. 461–476.

Bureau of Entomology Circular No. 92, Mites and Lice on Poultry; price, 5 cents.

Bureau of Entomology Circular No. 170, The Fowl Tick; price, 5 cents.

Yearbooks are usually to be found in local libraries or in the homes of some of the farmers.

ADDITIONAL COPIES
OF THIS PUBLICATION MAY BE PROCURED FROM
THE SUPERINTENDENT OF DOCUMENTS
GOVERNMENT PRINTING OFFICE
WASHINGTON, D. C.
AT
10 CENTS PER COPY
▽